SAE EDGE™
RESEARCH REPORT

Unsettled Issues Concerning Semi-Automated Vehicles: Safety and Human Interactions on the Road to Full Autonomy

Francesca M. Favarò, PhD
San Jose State University

EDGE DEVELOPMENT TEAM

Joseph Rodriguez, *San Jose State University*
Dr. Melissa Cefkin, *Nissan Research Center*
Dr. Alison Chaiken, *Peloton Technology*
Dr. Christian Janssen, *Utrecht University*
Angelos Lakrintis, *Strategy Analytics*
Dr. Mark Rosekind, *Zoox; former NHTSA; former NTSB*
Dr. Bernard C. Soriano, *California Department of Motor Vehicles*
Dr. Neville A. Stanton, *University of Southampton*
Dr. Ji Hyun Yang, *Kookmin University*

SAE INTERNATIONAL®
Warrendale, Pennsylvania, USA

About the Publisher

SAE International® is a global association of more than 128,000 engineers and related technical experts in the aerospace, automotive, and commercial-vehicle industries. Our core competencies are lifelong learning and voluntary consensus standards development. Visit sae.org

SAE EDGE™ Research Report Disclaimer

SAE EDGE™ Research Reports focus on topics that are dynamic, in which knowledge is incomplete, and which have yet to be standardized. They represent the collective wisdom of a group of experts and serve as a practical guide to the reader in understanding unsettled subject matter. They are not meant to provide a recommended practice or protocol. The experts have assembled as a community of practitioners to contribute and collectivize their thoughts and points of view; these are not the positions of the institutions or businesses with which they are affiliated, nor is one contributor's perspective advanced over other contributors. SAE EDGE™ Research Reports are the property of SAE International and SAE alone is responsible for their content.

About This Publication

SAE EDGE™ Research Reports provide state-of-the-art and state-of-the-industry examinations of the most significant topics in mobility engineering. SAE EDGE™ contributors are experts from research, academia, and industry who have come together to explore and define the most critical advancements, challenges, and future direction in areas such as vehicle automation, unmanned aircraft, Internet of Things and connectivity, cybersecurity, advanced propulsion, and advanced manufacturing.

Related Resources

SAE MOBILUS® Automated & Connected Knowledge Hub
https://saemobilus.sae.org/automated-connected

SAE Team

Frank Menchaca, Chief Product Officer
Michael Thompson, Director, Standards, Information and Research Publications
Monica Nogueira, Acquisitions Director
Beth Ellen Dibeler, Product Manager
William Kucinski, Managing Technical Editor

Printed in USA

ISSN 2640-3536
e-ISSN 2640-3544
ISBN 978-1-4686-0134-3

To purchase bulk quantities, please contact: SAE Customer Service

E-mail: CustomerService@sae.org
Phone: +1-877-606-7323 (*inside USA and Canada*)
+1-724-776-4970 (*outside USA*)
Fax: 724-776-0790

https://www.sae.org/publications/edge-research-reports

About the Editor

Dr. Favarò is an Assistant Professor in the Department of Aviation and Technology in the College of Engineering at San Jose State University (SJSU). Prior to joining SJSU, she earned a PhD and MS in Aerospace Engineering at the Georgia Institute of Technology, and MS and BS in Space Engineering at Politecnico di Milano, Italy. Dr. Favarò's research interests lie in the broad field of system safety and risk analysis, with an emphasis on system engineering concepts and the safe integration and embedding of new technologies and the consistent update of regulations and certification practices. In 2016, she funded the RiSA^{2}S lab, which deals with Risk and Safety Assessment of Autonomous Systems such as unmanned aircraft systems (UAS) and self-driving vehicles. In 2017, she became a research associate of the Mineta Transportation Institute and started collaborating as expert in the realm of autonomous vehicles (AV). Her interests are currently focused on the safe integration of autonomous systems within U.S. public roads as well as the National Airspace. She particularly focuses on bridging the gap between the technology world and the current regulatory panorama. Dr. Favarò has authored several journal publications and conference proceedings on a variety of topics, ranging from UAS mishaps to risk assessment and regulatory compliance for nuclear power plants. Dr. Favarò serves on the Board of Directors of the Aero Club of Northern California and acts as Secretary for the Community and Airport Partnership for Safe Operations, an aviation interest group that safeguards safe and efficient operations at the local Reid-Hillview Airport, San Jose. She is an Federal Aviation Administration certified Advanced Ground Instructor, a certified Remote Pilot for UAS commercial operations, and a solo-endorsed pilot.

contents

Unsettled Issues Concerning Semi-Automated Vehicles: Safety and Human Interactions on the Road to Full Autonomy

Abstract

Across the span of the SAE International-defined Levels of Driving Automation, human drivers occupy a diverse range of responsibilities and authority on the vehicle movement and the monitoring of the outside environment. From both a technological and a regulatory perspective, there is a gap that divides lower levels of automation (L1 through L3) and higher levels of automation (L4 and L5). For those vehicles that require the cooperation between a human driver and the autonomous technology, it is important to ascertain the safety consequences of such a design choice. It is also important to understand what the transition between automated driving and manual driving entails for the human driver, as well as for the surrounding traffic. This SAE EDGE™ Research Report investigates unsettled issues concerning what is commonly referred to as "semi-automation," including an overview of the role of human drivers, the quantification of the "transition-to-manual" problem, the role played by L3 toward full automation, and regulatory and moral considerations surrounding the deployment of these vehicles.

NOTE: SAE EDGE™ Research Reports are intended to identify and illuminate key issues in emerging, but still unsettled, technologies of interest to the mobility industry. The goal of SAE EDGE™ Research Reports is to stimulate discussion and work in the hope of promoting and speeding resolution of identified issues. SAE EDGE™ Research Reports are not intended to resolve the issues they identify or close any topic to further scrutiny.

FRANCESCA M. FAVARÒ, PhD

San Jose State University

Edge Development Team

Joseph Rodriguez, *San Jose State University*
Dr. Melissa Cefkin, *Nissan Research Center*
Dr. Alison Chaiken, *Peloton Technology*
Dr. Christian Janssen, *Utrecht University*
Angelos Lakrintis, *Strategy Analytics*
Dr. Mark Rosekind, *Zoox; former NHTSA; former NTSB*
Dr. Bernard C. Soriano, *California Department of Motor Vehicles*
Dr. Neville A. Stanton, *University of Southampton*
Dr. Ji Hyun Yang, *Kookmin University*

ISSN 2640-3536

Introduction: Premise (and Promise) behind Autonomous Vehicle Technology

Sought Advantages

Advanced Driver-Assistance Systems (ADAS) are becoming widespread features in many modern vehicles currently deployed on U.S. roads and on a global scale. One of the latest estimates for 2016 placed the production of vehicles equipped with ADAS at 25% of the total vehicles produced in the U.S., with an estimated growth of about 1.5% per year up to 2021 [68]. Such growth is fostered by the potential and promise of addressing important transportation issues, such as

i. The improvement of road safety, where human error is estimated to account for 94% of total accidents [65].
ii. The long-sought improvement of mobility for everyone, allowing differently abled people to access transportation more easily, and improving independence for people of all ages [53].
iii. The potential for a more sustainable transportation alternative, with facets ranging from fuel consumption optimization and efficiency [69] to low-carbon footprints for electric options [1] to more manageable parking arrangements and "green areas" for urban developments [2].
iv. The improvement of the commute experience, allowing the passenger to reallocate part of the commute time to tasks other than driving, and with the potential to shorten the commute once the car takes care of parking itself [2], as well as alleviating congestion. These factors would all contribute to more polite and relaxed users on the roadway.

Albeit noble in their target and clear in their scope, those goals might be (partially) met at various degrees, depending on the level of autonomy of the vehicle considered.

SAE International Levels of Autonomy and the Notion of "Semi-Autonomy"

Currently deployed and market-ready ADAS systems in the U.S. can mostly be mapped to Automation Levels 1 through 3 as defined by the SAE International standard for levels of autonomy [62]. SAE defined six levels of automation, ranging from Level 0 (no automation) to Level 5 (full unrestricted automation). The definition of the six levels (rows of Figure 1) is based on four factors (the four columns to the right of Figure 1) as follows:

i. The agent responsible for executing sustained lateral and longitudinal control of the Dynamic Driving Task (DDT): (human) driver and/or system automation
ii. The agent responsible for Object and Event Detection and Response (OEDR) of the DDT: (human) driver or system automation
iii. The agent responsible for serving as "backup" for the DDT after an automation failure: (human) driver or system automation
iv. The driving modes in which autonomous operations are allowed: either without one specific Operational Design Domain (ODD) (i.e., in all modes of operation, meaning unrestricted conditions) or within a limited ODD (i.e., meaning some pre-specified conditions, e.g., good visibility only)

Figure 1 depicts the schematic from the 2018-06 revision of the SAE J3016 standard. This revision aimed, among other things, at better defining the role of the human driver (if any) during the period in which the driving automation system is engaged [62]. The revision of the SAE standard brought with it more precise and novel terminology into the definition of the SAE levels of autonomy, and carried with it the novel term of "fallback-ready user" seen in Figure 1. This notion, which specifically refers to L3 technology only, is de facto responsible for the creation of a boundary between two "informal" categories of systems, which are currently at the center of public debates: semi-automated vehicles versus fully automated vehicles.

Levels 1 through 3 are regarded as "semi-automated" due to the fallback performance (or backup) of the DDT placed on the human driver. This points to a delicate interleaving between the driver and the system automation, with several implications on the need to verify the "user-readiness" to take over control in a timely and effective way - a problem at the center of the safety debate investigated in this report.

Conversely, Levels 4 and 5 are regarded as fully automated. Specifically, Levels 4 and 5 can be regarded as "restricted full autonomy" and "unrestricted full autonomy," respectively, with restrictions being placed on the ODD for the vehicle, and thus placed on the system capabilities in different driving modes and external conditions (e.g., fully autonomous vehicles that can only operate in daytime or under clear weather conditions or in particular locations/established paths).

Overview of the Current Market

Table 1 provides an overview of how the autonomous vehicles (AV) market is shaping up, with estimated timelines and levels of automation targeted by several major manufacturers. The global outlook estimate for the next 30 years shows an exponential growth for L3 and L4 vehicles (see Figure 2). The widespread fielding estimated for those levels thus elevates the importance of ascertaining the role that drivers will have within those vehicles in the future.

FIGURE 1. SAE levels of automation, SAE J3016 2018 revision.

Level	Name	Narrative definition	DDT: Sustained lateral and longitudinal vehicle motion control	DDT: OEDR	DDT fallback	ODD
Driver* performs part or all of the *DDT						
0	No driving automation	The performance by the *driver* of the entire *DDT*, even when enhanced by *active safety systems*.	*Driver*	*Driver*	*Driver*	n/a
1	Driver assistance	The *sustained* and *ODD*-specific execution by a *driving automation system* of either the *lateral* or the *longitudinal vehicle motion control* subtask of the DDT (but not both simultaneously) with the expectation that the *driver* performs the remainder of the *DDT*.	*Driver* and *system*	*Driver*	*Driver*	Limited
2	Partial driving automation	The *sustained* and *ODD*-specific execution by a *driving automation system* of both the *lateral* and *longitudinal vehicle motion control* subtasks of the *DDT* with the expectation that the *driver* completes the *OEDR* subtask and *supervises* the *driving automation system*.	***System***	*Driver*	*Driver*	Limited
***ADS* ("*System*") performs the entire *DDT* (while engaged)**						
3	Conditional driving automation	The *sustained* and *ODD*-specific performance by an *ADS* of the entire DDT with the expectation that the *DDT fallback-ready user* is *receptive* to *ADS*-issued *requests to intervene*, as well as to *DDT performance-relevant system failures* in other *vehicle* systems, and will respond appropriately.	*System*	***System***	*Fallback-ready user (becomes the driver during fallback)*	Limited
4	High driving automation	The *sustained* and *ODD*-specific performance by an *ADS* of the entire *DDT* and *DDT fallback* without any expectation that a *user* will respond to a *request to intervene*.	*System*	*System*	***System***	Limited
5	Full driving automation	The *sustained* and unconditional (i.e., not *ODD*-specific) performance by an *ADS* of the entire *DDT* and *DDT fallback* without any expectation that a *user* will respond to a *request to intervene*.	*System*	*System*	*System*	**Unlimited**

Unsettled Issues Concerning Semi-Automated Vehicles: Safety and Human Interactions on the Road to Full Autonomy

At the heart of the safety debate presented in the Introduction is the role of human drivers in their interaction with an ADAS-equipped vehicle that can be subject to brittle automation. In semi-automated vehicles, a human driver shares authority of the DDT with the automated driving system (ADS) that acts as the "brain" of the vehicle, and furthermore serves as backup whenever the system disengages from the automated mode. Disengagements can be due to a variety of factors (e.g., due to insufficient information regarding the context of a particular situation or due to particular design restrictions on the ODD), and regardless of their trigger modality, they reveal the delicate nature of the transition of control authority from the system to the human driver, often termed the "transition-to-manual" problem. This report thus investigates the main ingredients that inform the safety debate on whether it is possible to ensure that human drivers have enough time to react and respond

TABLE 1. Overview of testing, deployed, and targeted levels of automation for major companies testing in North America (not meant to be exhaustive). The data points were estimated based on multiple media outlets, from manufacturers' websites, public statements, and through data obtained thanks to Strategy Analytics (see table notes for references).

Manufacturer	Market deployed level	Testing level	Target level	Target year for North America testing	Notes
Waymo	4	4	4	2020+	(a), (b)
Audi Auto Group	2-3	N/A	N/A	N/A	(c)
BMW	2	3/4	3/4	2021	(c)
Volkswagen	1	2	2	2019	(c)
Porsche	1	N/A	N/A	N/A	(c)
Toyota	1	4	4	2020+	(c)
Renault	2	4	4	2022-2025	(c)
Nissan	2	2	2	2019	(c)
Hyundai	1	4	4	2022	(c)
Ford	1	4	4	2020	(c)
Fiat Chrysler Auto	1	4 w/Waymo	4	2020	(c)
Mercedes-Benz	2	4	4	2020	(c)
Peugeot S.A.	1	2	2	2019	(c)
Volvo	2	4	4	2020	(c)
Tesla	2+	4	4	2021	(c)
Jaguar Land Rover	2	2+	2+	2020	(c)
General Motors	2	3	3	2019	(c)
Honda	1	3	3	2020	(d)
Subaru	1	2	3+	2024	(e)
Zoox	*	4/5	4/5	2020-2021	(f)
Baidu		2	4	2020	(g)

* Approved to operate autonomous ride service.

(a) https://www.reuters.com/article/us-waymo-selfdriving/waymo-picks-detroit-factory-for-self-driving-fleet-to-be-operational-by-mid-2019-idUSKCN1RZ1IZ.

(b) https://techcrunch.com/2019/07/02/waymo-is-now-allowed-to-transport-passengers-in-its-self-driving-vehicles-on-california-roads/.

(c) Angelos Lakrintis for Strategy Analytics, OEM Autonomous Vehicle Strategies, June 2019, https://www.strategyanalytics.com/access-services/automotive/autonomous-vehicles/reports/report-detail/oem-autonomous-vehicle-strategies-2019.

(d) https://www.cnet.com/roadshow/news/honda-self-driving-car-technology-japan/.

(e) https://www.cnet.com/roadshow/news/subaru-hybrid-suv-autonomous-plan/.

(f) https://zoox.com/wp-content/uploads/2018/12/Safety_Report_12Dec2018.pdf.

(g) https//usa.baidu.com/adu/.

effectively to the request to control the vehicle. Specifically, this SAE EDGE™ Research Report considers the following areas to be unsettled domains related to semi-automated vehicles, as marked by an apparent lack of common understanding regarding key aspects of each:

- The role of the human driver
- The "transition-to-manual" problem and the metrics employed to quantify and qualify the takeover performance
- The role of L3 (if any) on the progression to full autonomy
- The regulatory and moral challenges associated with L3 vehicles

The Role of the Human Driver[1]

A substantial body of literature started to surface in the 1970s to study the interaction between human operators and automated systems (see, for example, [48, 64], and [5, 73] for more recent efforts). On-road studies and simulator studies also surfaced to better understand the notion of an "intermittent operator" when shared authority is considered [11]. What has been contentious, however, is the definition of the "role of the driver" and task allocation between the operator and

[1] This section heavily relies on SAE J3016 terminology. Readers unfamiliar which such standard should consider reviewing the information therein first.

FIGURE 2. Global new car market outlook in terms of sales volume.

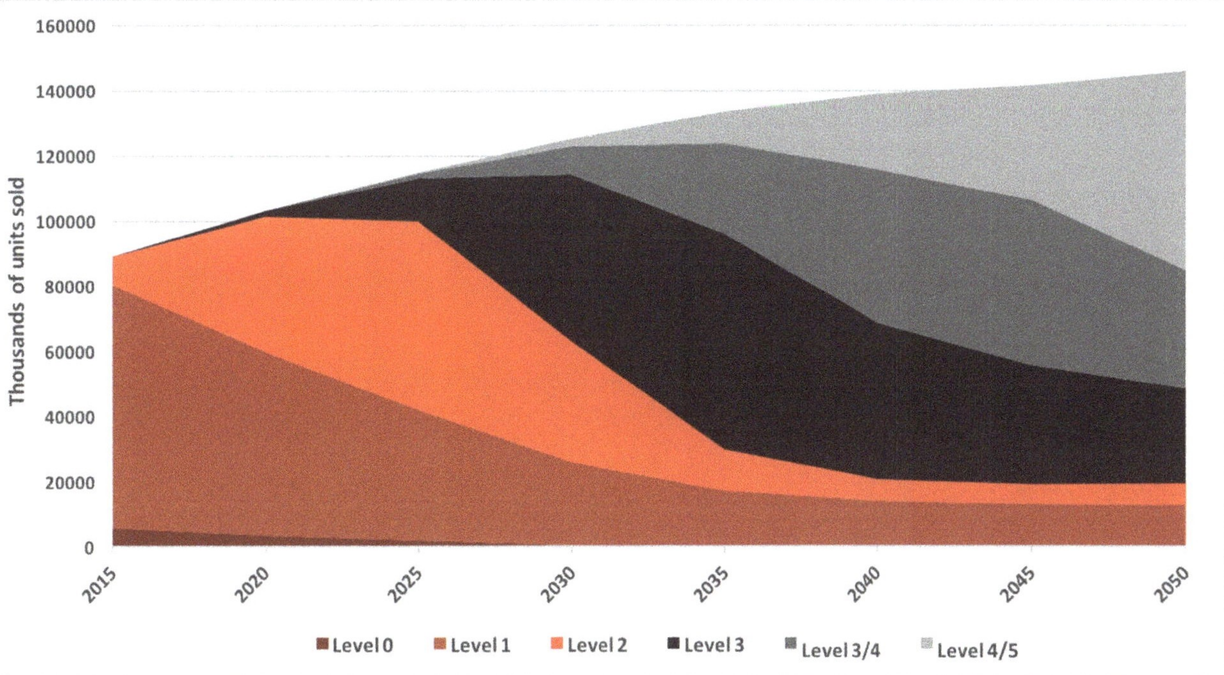

the system under different levels of autonomy [6].[2] The new revision of SAE J3016 undertook a major effort in trying to better define such a role. Yet, agreement in the academic community and across the different disciplines of engineering, psychology, and human factors and ergonomics is that clarifications on key terminology are still needed. This section is aimed at summarizing some of the current contentious points on the role of human drivers and their functions.

Driver's Function within the DDT

Revision 2018-06 of the SAE J3016 standard brought a considerable change within the factors that help discern the six levels of automation. In particular, this revision brought further refinement to the definition of the DDT into the subcategories of

1. Sustained lateral and longitudinal vehicle motion control
2. OEDR
3. DDT fallback (i.e., backup after failure, term fully defined later)

[2] A recent study suggested new human-centered levels of automation based on tasks that drivers could undertake in different levels. The new levels when then compare to the SAE J3016 levels, highlighting inconsistencies on the role of the human driver. A key conclusion was that SAE levels are based off the technological capabilities of the vehicle and are thus difficult to interpret from the driver's standpoint [78].

The second factor listed above replaced the 2016 version of a function called "monitoring of the external environment." Note that the 2018-06 new factor now combines both the detection and the response within the same category. The merge of the two different functionalities reduces the refinement with which the role of the human driver can be described under the current standard, as explained next.

From a human factors perspective, three basic skills (i.e., cognitive, perceptual, and motor) are associated with different driving functions [7, 66] as indicated in Figure 3. The functions indicated in Figure 3 show a consistent matching to those that any automated vehicle must manage, which, in conventional driving, fall within the purview of the human operator. Those functions, along with their associated questions, are depicted in Figure 4 and are summarized as follows:

- Mapping and Localization: answering the fundamental question "Where am I?"
- Perception: answering the fundamental question "Who/what is around me?"
- Prediction: answering the fundamental question "What is everybody else/everything else going to do?"
- Planning: answering the fundamental question "What am I going to do?"
- Execution: performing the plan while continuing to carry out the previous functions and adjusting the plan as necessary

FIGURE 3. **Simple schematic of basic skills and associated driving functions.**

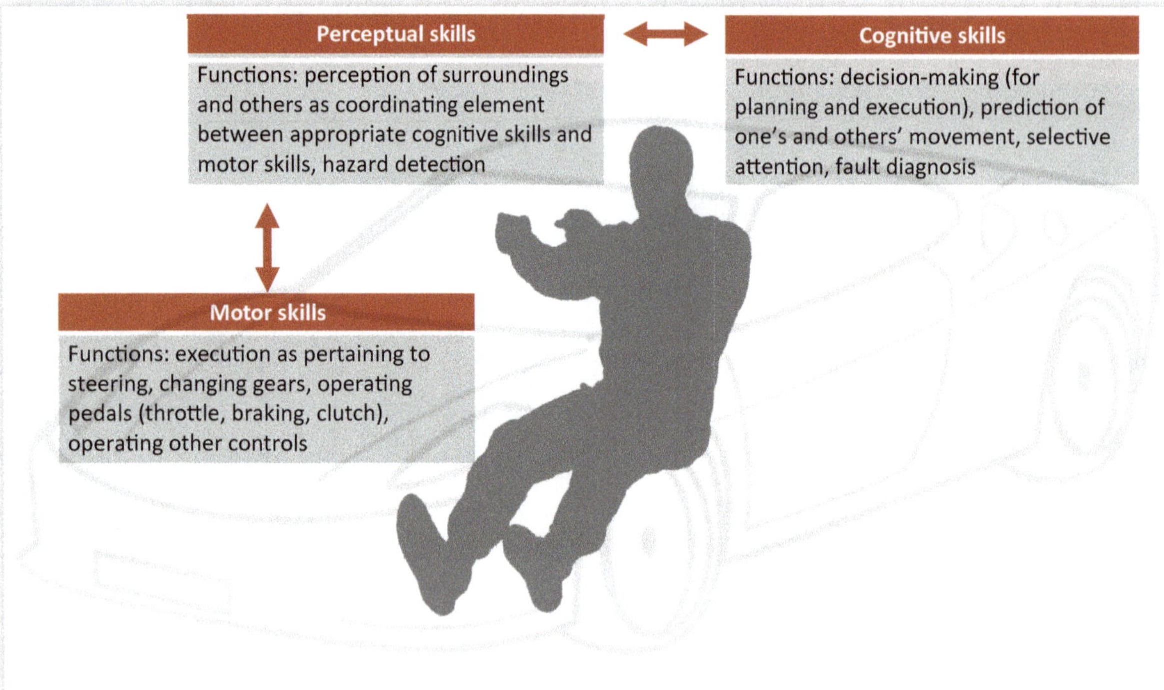

FIGURE 4. **Schematic of necessary driving functions (not meant as an actual control loop).**

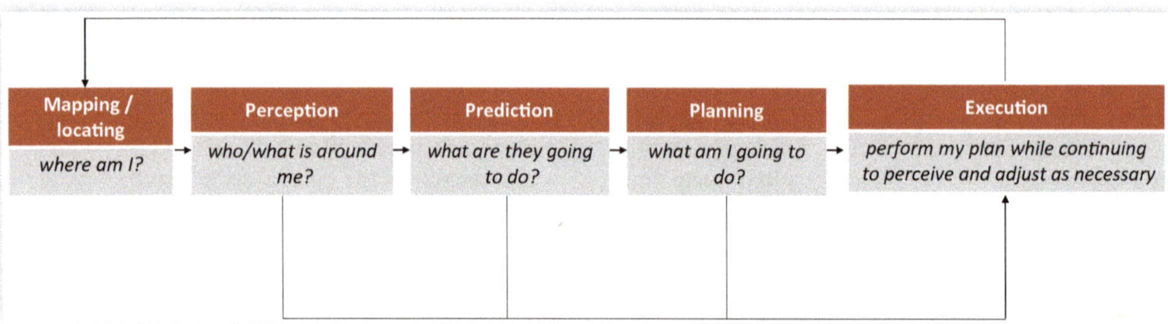

When carried out by a human operator, those functions can all be traced back to cognitive, perceptual, or motor skills as identified in Figure 3.

From this standpoint, the aggregation of the "detection" and "response" functions within the second factor of SAE J3016 can hamper the proper definition and/or allocations of skills necessary for the accomplishment of such functions on the part of the agent responsible for them. Specifically, the "detection" function falls within the perceptual skills of Figure 3, while the "response" functions fall within the motor skills of Figure 3. Moreover, "detection" and "response" are separate functions that could be undertaken by different agents (e.g., ADS - system detection and human driver response), so that their combination reduces the granularity of the driver's role description within the levels of autonomy definition.

From a more holistic perspective on a driver's role within an automated vehicle, SAE J3016 revision 2018-06 opened up the possibility of a driver fallback response following ADS disengagements in both L4 and L5 vehicles. A sample situation included within the standard is depicted in Figure 5, which accounts for the fact that a passenger could become a driver, as needed. Note that SAE J3016 allows such transition from the role of passenger to the role of driver (and vice versa) for both L4 and L5, based on a passenger's possibility to request an ADS disengagement and/or engagement in the first place (see Table 2 within the standard, pp. 21 and following).

The case of Figure 5 can be considered straightforward from a system engineering perspective. An L4 vehicle is approaching an operational condition beyond the intended ODD; two responses are possible: having the

FIGURE 5. Sample use case sequence at Level 4 from SAE J3016, showing ADS engaged with ODD exit, which does not prevent continued DDT performance by an available human user [62].

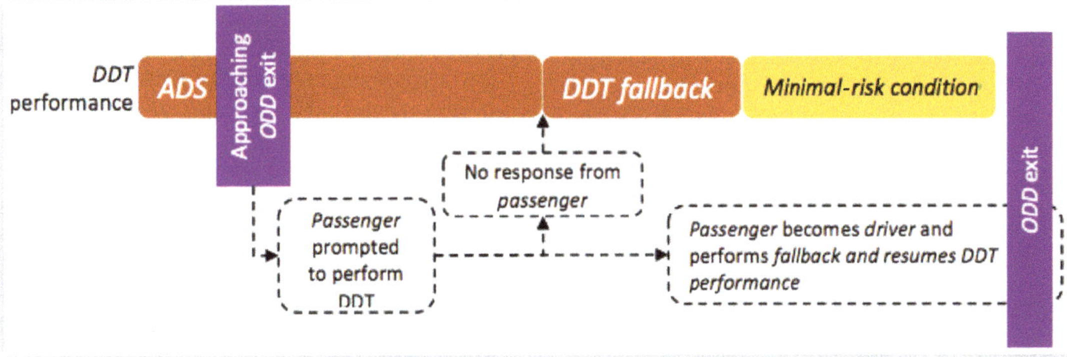

passenger transition to the role of driver and resume the DDT manually or having the vehicle autonomously reach a minimal risk condition. Sample use cases for L5 vehicles are instead not provided within the standard. Within the framework laid forward by the new standard's revision, a passenger within a fully automated vehicle could assume the role of driver when the vehicle is incapable of reaching a minimal-risk condition (which should only occur when the vehicle is operating beyond its intended ODD due to off-nominal emergency situations). The possibility of human driver takeover implies the existence of a control interface (e.g., steering wheel and pedals) that would permit the occupant to respond to a "request to intervene" and assume authority of the DDT of a L4/L5 vehicle. A passenger must therefore consider the possibility of acting as a backup driver also in a L4/L5 vehicle. In addition, the passenger also has the authority to voluntarily disengage the ADS (see Table 2, pp. 22-23 for L4 and L5 roles within the SAE J3016 standard). Thus, unless ADS-assigned authority is designed to override such input also in failure scenarios, a driver could regain control of the vehicle. History taught us that software overrides need to be carefully considered when interleaving with automation is involved, and the possibility of a switch from passengers to driver within L4/L5 must then be carefully assessed, and verbiage within the standard possibly revised in those regards. The possibility for the human user to act as "backup," in turn, implies the need of a "fallback-ready user" within the full-automation context if beyond the ODD of the vehicle. In other words, the "DDT fallback" role of the driver would no longer be tied to the specific level of automation, but to whether the vehicle is operating within or outside the limits imposed by the nominal ODD for the vehicle. This nuance should not be underplayed and requires careful consideration from a standards perspective, as it carries with it requirements on the interface design equipped within the vehicle and requirements on the agent to whom authority is assigned together with design choices on possible input overrides. But what does it mean to be "fallback-ready"?

Fallback, Monitoring, and Supervision

According to SAE J3016 2018-06, at levels L3 through L5, the ADS is responsible for monitoring its own performance on the entire DDT. A "supervisory" role has to be maintained by the driver only for L1 and L2, as conveyed by the following definition:

> *Supervise definition: "The driver activities, performed while operating a vehicle with an engaged level 1 or 2 driving automation system feature, to monitor that feature's performance, respond to inappropriate actions taken by the feature, and to otherwise complete the DDT." SAE J3016, 2018-06*

The standard defines "fallback" as "the response by the user to either perform the DDT or achieve a minimal risk condition after occurrence of a DDT performance-relevant system failure(s) or upon ODD exit, or the response by an ADS to achieve minimal risk condition, given the same circumstances." Moreover, the term "fallback-ready" is tied to the notion of a user that is "able to operate the vehicle" and "receptive to ADS-issued requests to intervene and to evident DDT performance-relevant system failures." Receptivity is "an aspect of consciousness characterized by a person's ability to reliably and appropriately focus his/her attention in response to a stimulus" and where

> *In level 3 driving automation, a DDT fallback-ready user is considered to be receptive to a request to intervene and/or to an evident vehicle system failure, whether or not the ADS issues a request to intervene as a result of such a vehicle system failure. SAE J3016 2018-06*

A key problem thus becomes that of quantifying and qualifying how a user can be considered "receptive." This consideration is at the heart of the discussion within the next section on "Transition to Manual." Yet the previous section discussion revolved around the fact that verbiage currently included within the SAE J3016 standard leaves open the possibility of a human driver "fallback" at higher levels of automation (with the system being the primary designated responsible agent per Figure 1), with consequences that need to be better discussed and defined.

Recommendations

Areas of improvement related to the terminology and the overarching context for the definition of the role of the driver for different aspects of the DDT have been identified and can be summarized in the following recommendations:

1. Further discussion is needed on the aggregation of the "detection" and "response" function for external objects and events as proposed in the 2018-06 revision of SAE J3016 in its role of differentiating among levels of autonomy, given that the two functions require different skills (associated to different realms of human factors), and could be carried out by different agents.
2. The relation between the role of the human driver and the ODD has to be further explored. In particular, the transition from the status of "passenger" to that of "driver" could be associated with transitions beyond the limits of the ODD for L4 technology. L5, given its theoretically unlimited ODD, would require a separate consideration to assess under which conditions the "passenger" can request an ADS disengagement (if at all). This would be tied to the problem of assigning authority (with possible override of the driver input) of the fallback in emergency situations.
3. Based on the two previous considerations, a revision of the definition of the term "fallback-ready" may need to be considered within the upcoming revision of the J3016 joint standard between SAE and ISO. While the role of "ADS issued requests" is analyzed later in this report, at this stage the recommendation brought forward is to clarify the reasoning behind why such term is tied to the level of automation, since many of the ingredients within the definition of "fallback-ready" and "reception" seem general enough to relate to both L3 and L4 as well as to the specific ODD limitations for the vehicle (e.g., considering for instance a fallback-ready user whenever the vehicle finds itself beyond the ODD, regardless of it being L3 or L4).

Transition to Manual

Beyond the notion of "fallback," which is starkly user centered as defined by SAE J3016 ("response by the user"), the notion of "transition of control authority" and the ADS-centered notion of "disengagement" also must be defined. The California Department of Motor Vehicles was the first state agency to provide a definition for the technical term "disengagement" and to mandate yearly reporting of disengagements' records from those manufacturers that wished to test their systems on public roads. A disengagement is defined as a transition between two different operating modes of the vehicle: from the "automated mode" to the "conventional mode" [16, 17]. Furthermore, the automated mode is defined as the "status of vehicle operation where technology that is a combination of hardware and software, remote and/or on-board, performs the dynamic driving task," while the conventional mode is the "status of the vehicle when it is under the active physical control of a natural person sitting in the driver's seat operating or driving the vehicle with the autonomous technology disengaged" [16, 17]. In simple terms, a disengagement happens whenever the vehicle control authority shifts from the automation that acts as "brain" of the vehicle to the human driver. The transition in control authority could be due to a number of reasons both inside or outside of the vehicle, and can be initiated both manually by the user (i.e., operator's request), or can be automated (e.g., a sudden failure of the automated software or hardware, or a situation the vehicle was not designed to handle in its intended ODD) [30]. In either case, after a disengagement happens, the driver is responsible for the safe operation of the vehicle, and must continually maintain situational awareness of their surroundings and of the vehicle's performance through the functions highlighted in Figure 4.

From a human factors' perspective, this problem has been studied in a host of different systems, starting within the aerospace industry, and it is often referred to as the "transition-to-manual" problem. These types of control authority transitions can be safety-critical, depending on the specific scenario in which they present themselves (e.g., vicinity of other cars/people/property, closeness to shoulders/sidewalks/construction areas).

The research community thus deemed it important to quantify how drivers respond to such situation and assess their ability to effectively regain control of the vehicle. Approaches to quantify and qualify what a "successful" takeover performance is have entertained the research community for more than 10 years. They are characterized by the need for better standardization and for a guiding effort that could steer the research paths in a common direction, as reviewed next.

Takeover Performance Metrics

The first point raised by collaborators of this SAE EDGE™ Research Report is that the academic community is contributing to the proliferation of metrics aimed at quantifying the takeover performance. However, not all those metrics can be directly applied in an operational or regulatory context. A point of discussion was thus drafting key requirements that a metric should possess. This section thus provides a classification of the most commonly employed metrics within the AV community, and presents the requirements for an effective metric.

What is an effective metric?

When trying to quantify the "effectiveness" of a switch in control authority of a vehicle, the collaborators identified four fundamental requirements that an ideal metric would meet:

R1. It needs to be informative: This quality is related to the "why" and the reason for measuring a particular variable. "Informative" means that it must be possible to abstract information useful enough to make assertions about the metric.

R2. It needs to be measurable: Although not necessarily numeric (in which case the metric would become quantifiable), a metric (either discrete or continuous) needs to have specific categories or set of values that allow the user to differentiate among different instances and measurements.

R3. It needs to be rankable and/or own associated thresholds: It is important to be able to assign bounds or thresholds associated with "safe" and/or "unsafe" states. Bounds can be further refined to account for different safety margins and cautionary states. If bounds cannot be determined, it is important that the metric be at least rankable, so that it would be possible to assess which value between two different measurements is closer to the intended threshold.

R4. It needs to carry operational bearing: A safety threshold alone is not sufficient if it does not carry practical value within an operational context. Moreover, the regulatory counterpart of a metric should also be considered. Regulations are often provided in the form of a limit (e.g., speed limit) and so are certifications (e.g., "probability of failure less than"), so that having an operational bearing is of paramount importance for widespread use beyond the academic environment.

Figure 6 provides an example of how all criteria are met in the case of Blood Alcohol Concentration (BAC) measurements that are used to assess whether a driver is Driving Under the Influence (DUI).

The majority of metrics currently employed to quantify takeover performances meet requirements R1 and R2, but not all meet requirement R3, and it is tough to map each metric to requirement R4 (i.e., creativity and ingenuity are needed to answer the questions "How would I use this metric in an operational context? Can the metric bear regulatory value?" with no straightforward resolution).

From a practical standpoint, there are additional requirements that would be important to ensure, but those are not at the center of the foundational debate. These requirements are more tied to an optimization problem

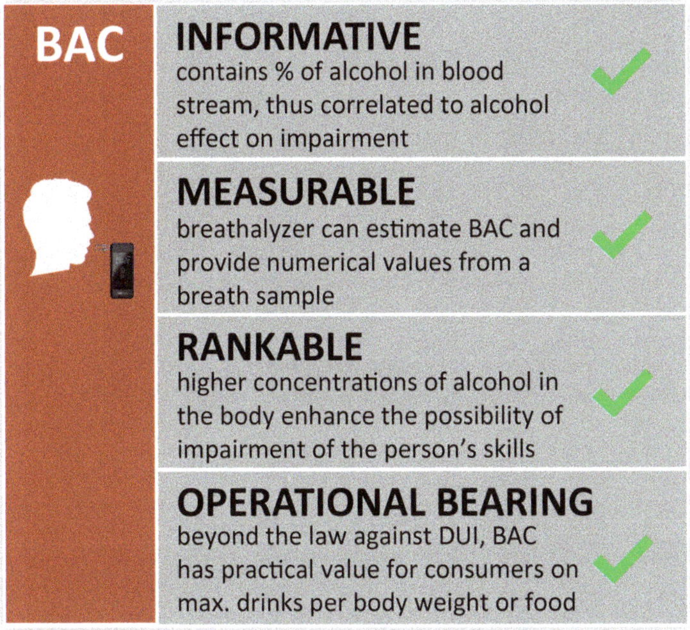

FIGURE 6. Example of a metric that meets the criteria for effectiveness: blood alcohol concentration (BAC) used for DUI assessment.

related to technological barriers rather than to their essence and definition, and are not considered within this discussion at this stage. Among those are

1. The need to be easily scalable
2. The need to be measurable in (close to) real time

Metrics Overview and Classification Although no agreed-upon classification exists for takeover performance metrics, consensus among the experts was that at a first layer of understanding there are two broad categories of metrics:

- Those that are driver centric: tied to the performance of the human driver in relation to the functions described in Figure 4
- Those that are vehicle centric: tied to trajectory-based variables pertaining to the movement of the vehicle

Table 2 provides an overview of the most commonly employed metrics found in the literature, along with key

TABLE 2. Summary of generalized common metrics employed in literature for takeover performance, with sample references for use cases.

Metric	Definition	Sample references
Time-to-input	Time to first recorded input or time to specific input (e.g., steering). Usually amongst braking, throttle, and steering. Often employed as a proxy for "time to takeover."	[12, 41]
Time-to-action	Time to initiation or time to completion of specific actions. Can also account for time-related gaze tracking metrics. Example: time to eye gaze on the road, or time to manual disengagement of ADS. Can also be employed when interleaving is considered, and the ADS is repeatedly turned on/off. Example: time to re-engage ADS after first takeover.	[35, 80]
Time-to-condition	Time until a pre-specified condition is met. Can be tied to both the operator and the vehicle. Example: time to return within lane or time to return to heart rate within specified bounds	[25]
EEG	Electroencephalogram measurement: monitoring and recording of electrical activity within the brain. Can be related to emotional stress, attentiveness, drowsiness, vigilance.	[39, 71, 75]
EKG/ECG	Electrocardiogram measurement: under this category all hear rate monitoring and blood pressure monitoring metrics can also be subsumed, such as BVP (blood volume pressure) and HRV (hear rate variability). Typically used more for comfort-related metrics and drowsiness.	[37, 44, 56]
GSR	Galvanic Skin Response: measurement of the electrodermal activity. Typically related to emotional stress.	[59]
GD	Gaze Dispersion: absolute (i.e., not time related) eye tracking metric to quantify deviation of gaze position within a particular direction (e.g., horizontal or vertical). Typically measured in degrees. Can also include indications of gaze fixation and movement across different objects.	[36, 45]
TLX/workload	Task Load Index: subjective workload assessment to quantify mental/physical/temporal demand, perceived levels of effort required to complete a task, frustration experienced, and estimate of self-performance.	[10, 22, 27, 63]
Recollection	Subjective measure related to recollection of details (e.g., speed of the vehicle, time within simulation, external objects) to assess situational awareness.	[49, 61]
Free response metrics	In this category fall all other subjective measures for self-assessment. Examples include indications of levels of trust and comfort during the takeover maneuver.	[24, 60]
Speed stability	Analysis of speed behavior within a given time horizon to highlight erratic behavior in speed management, and damping of oscillatory behavior.	* [34]
Position stability	Analysis of lateral offset behavior (typically from the centerline of a lane) within a given time horizons to highlight erratic behavior in position management, and damping of oscillatory behavior.	[26, 47]
Absolute - lateral offset	Any absolute (i.e., not time related) measure related to vehicle lateral offset, such as maximum/average or integral offset. Also includes measures of dispersions such as standard deviations. Can be tied to absolute measures of the steering wheel angle and/or vehicle heading too.	[20, 26, 30, 41]
Absolute - speed**	Any absolute (i.e., not time related) measure related to vehicle speed, such as maximum/average or integral velocity. Also includes measures of dispersions such as standard deviations. Can be tied to absolute measures of the throttle/braking pedal compression too.	[21]
Stopping distance	Measurement of distance to come to partial stop or full stop (different threshold can be assigned), typically as pertaining to achieving a minimal-risk condition.	* [34]

* Note: These metrics are commonly employed for the analysis of control algorithms for full-automation.
** Acceleration-related metrics can also be considered, although typically less intuitive when vehicle based, or tied to the already considered throttle usage when user based.

FIGURE 7. Classification of identified metrics (not meant to be exhaustive).

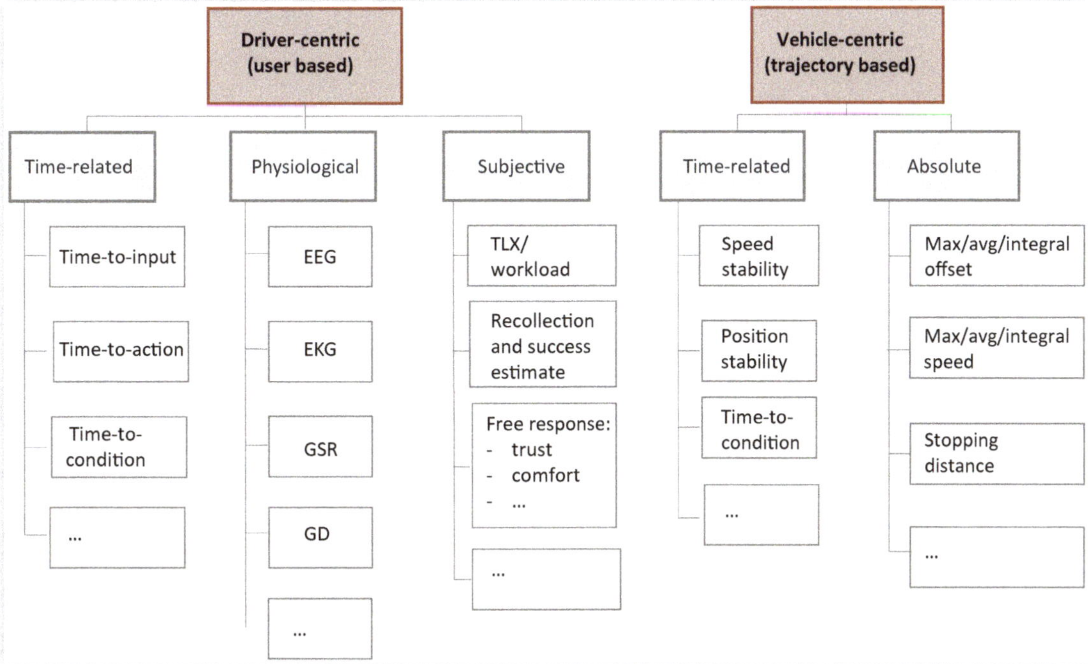

references for the interested reader to explore sample uses. The definitions provided in Table 2 are generalizations of the uses that can be found in the literature. Note that no agreed-upon definitions were found, and collaborators pointed to the fact that each work typically starts with redefining the metrics employed in relation to the specific (often limited) scope of the work carried out.

The metrics generalized in Table 2 can be further classified according to the two-tiered structure presented in Figure 7. The proposed taxonomy is not meant to be exhaustive and provides a starting point to further the discussion on takeover performance metrics, beyond the current status quo. Also note that all the metrics here considered are "pure," as opposed to aggregate measures that can be obtained once two metrics are combined within a new one. For examples of aggregate measures, the interested reader is referred to the examples contained in [14, 50], and [77].

Unsettled Issues and Recommendations

The following concerns were identified following the discussions among the collaborators:

1. **Need to consistently define time-related driver-centric measures**: As mentioned before, many of the metrics employed in takeover studies get defined for the specific study at hand, and there is a lack of consensus on the exact terminology as well as on the measurement techniques employed. While the problem is recognized by many [33], efforts to tackle cohesive definitions are lacking and there is a need for standardization. This problem is refined into two particular concerns:

   - Pinpointing the "time to takeover (TTO)": TTO is a metric widely employed in the literature included in Table 2. Yet, authors mean different things, some of which can be traced back to time-to-input, time-to-action, or time-to-condition. Consensus has first to be met on why this metric is assessed; the information sought is indeed what should drive a standardization on the actual method of measurement for such quantity. Standards in those regards should also be wary of the distinction between the notion of "reaction time" and "response time," two terms often used interchangeably, but that account for different levels of awareness and human activity [76]. Figure 8 showcases examples of various TTOs found in the literature, to ensure the community remains mindful when comparing data from different studies. Even if consensus could not be met for a general definition of TTO, the requirement should be set that, at a minimum, each study includes a clear description of the methodology employed to measure TTOs as well as the specific thresholds employed to define such time, which unfortunately is not always the case. The issue of defining thresholds is also tied to ensuring

FIGURE 8. **Sample use cases for TTO computation. The TTO changes depending on the experimental setup employed for measuring this quantity, hampering comparison if scenario standardization is not employed. CG = center of gravity.**

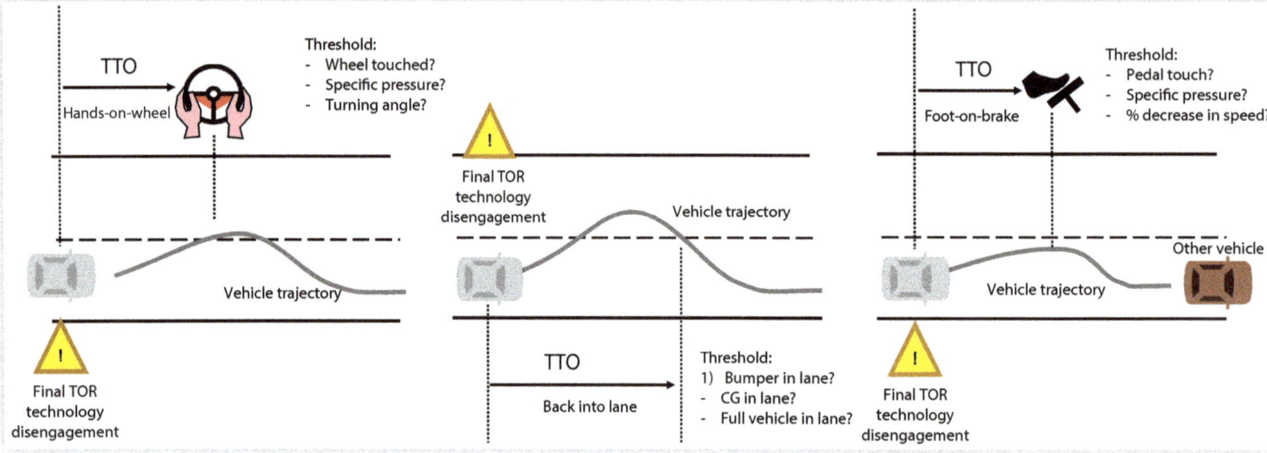

that the TTO is capturing a voluntary response from the user. At this time there is no consensus on how/whether TTOs should discern between an involuntary reflex and a voluntary response; therefore a deeper discussion on the rationale behind such metric is needed. Figure 8 brings forward some hypothetical questions within the definition of thresholds for each of the sample TTOs, which should be considered whenever assessing the boundary between involuntary and voluntary response.

- Understanding the type of disengagement investigated: two families of disengagement scenarios can be identified, and while time-related metrics can be assessed for both, no confusion should be made in comparing results across the two categories. Specifically, disengagement scenarios are divided into structured versus unstructured. The differentiating factor lies in whether a pre-warning to the final takeover request (TOR) is issued or not. A pre-warning constitutes a "heads-up" that the vehicle provides to the driver, prepare them to resume control in a short while, or in other words, to "prime the driver" and notify that the actual disengagement is about to happen soon [40, 70]. A pre-warning is also technically referred to as a "staged-alert" [11]. The final TOR is instead the takeover request that happens at the actual time of disengagement. A TOR is also provided in the form of a notification issued by the vehicle, so that the term "warning" can sometimes be misunderstood in the context of semi-automated vehicles' disengagements. The key difference is the timing of such a warning, with a final TOR happening at the time where the ADS is no longer in control of the vehicle, and pre-warnings happening prior to that, when automation is still engaged. A disengagement can have both a preparatory alert and a final TOR so that the two types of warnings are not exclusive [33]. Based on this distinction, we can speak of "structured disengagements," where a prior warning is issued to prime the driver before the disengagement takes place, and "unstructured disengagements," where no prior warning is present. This consideration is important when assessing and comparing time-related metrics, as visual cues like road obstructions and lane closures can possibly spike the attentiveness of the driver and improve response times compared to those obtained after a purely unstructured and "unmotivated" disengagement (such as a sudden system failure). The distinction motivates addition of further details to the scenarios in which TTOs are assessed from a research perspective - a common trait within the regulatory context, where safety scenarios are assigned, for instance, to test behavioral competencies of the vehicle [54, 55]. Furthermore, drivers' takeover performance in terms of trajectory stability after structured disengagement was shown to dramatically depend on the timing at which the staged alert is provided [40, 49, 70], with, generally speaking, better responses shown when a longer time for priming is provided and alerts repeated up until the point of actual disengagement. Two types of structured disengagement then must be differentiated: those that can be identified at the beginning of a trip (e.g., due to geofencing and location restrictions through the path panned for the vehicle) and those that get identified during the trip (e.g., weather degradation below the minima handled by the vehicle). The distinction between the two types is important because it can affect the driver response; moreover, the driver could also have the opportunity to decide to reroute the vehicle to limit the number of takeovers in a trip for those that are anticipated from the get-go.

2. **Need to further the dialogue among different stakeholders:** The proliferation of metrics employed within the academic environment is a symptom of a missing connection between the theoretical research world and the more practical world of regulators and manufacturers. A careful analysis on the effectiveness of each metric is needed (e.g., by checking that requirements such as those proposed here - R1 to R4 - are met). This analysis would ensure that the solutions proposed carry operational bearing, while remaining intuitive. This conversation can also further the definition of standards for experimental scenarios and apparatuses for metrics' measurements discussed in the previous point. Furthermore, structured conversations across stakeholders could also ensure that all parties will share the same terminology moving forward, considering that we are still at the early stages of adoption of ADS technology. While no consensus on which metrics could be best to describe the takeover performance (and all those included in Table 2 are still being actively considered), collaborators for this report agreed that so far little to no operational bearing has been embedded within them.
3. **Clarifying the role of independent variables**: From a technical standpoint, the metrics discussed in the previous section are considered "dependant variables." Dependent variables, as the name implies, are assumed to depend on a specific set of other variables, termed "independent." A host of independent variables has been investigated within the literature, ranging from road geometry to speed at the time of disengagement, age of the driver, driver experience, and many more (a review of targeted independent variables is included in [[28], under review] and a good sample of uses can be found in [11]). Similar to what was discussed for the experimental scenarios for TTO testing, the stakeholders could benefit from a common list of independent variables of interest. Each study would then provide a specific mapping of what variables are targeted out of the list of the common and standardized set, and steer research paths toward a common direction. A standardized set of independent variables should be agreed upon beyond the academic environment, to include all stakeholders.
4. **Unify research paths in relation to human-machine interfaces (HMI)**: HMI serve both the role of notifying the human operator about any change within the ADS (output role) and the mechanism through which the operator can provide inputs to the vehicle. A substantial body of literature exists that reviews how different HMI designs impact takeover performance (see, for instance, [3, 4, 8, 9, 11, 27, 49]). Cross-communication within the community that is producing the bulk of state-of-the-art data is needed to ensure that research is moving more consistently in a common direction. Furthermore, the role of HMI for both input and output interfaces within the revised standard SAE J3016 2018-06 is somewhat contentious and could be further clarified:
   - In its output role, SAE J3016 2018-06 notes that the ADS of L3 vehicles may not provide a warning to the operator of an upcoming disengagement in the case of "evident failures" (see the definition of "receptive" included before, where the standard specifies that the user should be ready to take over control in the case of evident failures "*whether or not the ADS issues a request to intervene as a result of such a vehicle system failure*," pp.14). Research into how a user perceives "evident failures" without the issuance of a TOR should then be tackled, while a definition of what is considered "evident" should be provided.
   - In its input role, SAE J3016 2018-06 raises the issue of a "not drivable" vehicle when it mentions that requests to intervene are defined as a "*notification by an ADS to a fallback-ready user indicating that s/he should promptly perform the DDT fallback, which may entail resuming manual operation of the vehicle (i.e., becoming a driver again), or achieving a minimal-risk condition if the vehicle is not drivable.*" The status of "not drivable" needs further clarification and seems to imply the absence or failure of input HMIs, another point that should be discussed.

Beyond Unsettled: Further Areas of Research The recommendations identified in the previous section stem from a considerable amount of published research and state-of-the-art discussions among policymakers, manufacturers, and academia. Those are, in a way, known problems that have made piecemeal appearances over the last two years, for which no clear solution has been identified. There are, however, outstanding research questions that the collaborators of this SAE EDGE™ Research Report felt deserving of further attention from the community in the future. Those include

1. **The importance of longitudinal studies and behavioral adaptation**: The majority of ongoing research covers limited time spans (in the range of a few minutes to hours, and few studies over few days [42]). There is a need for both simulated and naturalistic research to understand the effect of

long-term use of semi-automated systems. Specifically, the following areas of interests were identified:

- Drowsiness in sustained operations: Driver distraction and fatigue related to sleep deprivation have been studies for over three decades. Within the aviation sector, the existence of micro-sleeps and decreased brain activity even during critical stages of approach and landings have been observed and analyzed since the 1970s [15, 58, 74]. Within a semi-automated vehicle, concerns for drowsiness can be heightened by the fact that, except in commercial vehicles, the user is not a trained operator, and by the shorter timescales for collision avoidance with surrounding traffic. Over long periods of times, additional factors kick in (reviewed next) that can further contribute to a person giving in to drowsy states.
- Complacency and de-skilling: When considering sustained use from weeks to years, users will likely experience behavioral adaptation. Within a five-day study, increasing trends for trust were already witnessed, even after unforeseen situations that required emergency takeover [42]. Beyond complacency, there is also an issue of "skills retention" (once more a problem studied in aviation [19]) when automation is engaged often over extended periods of time. Tied to the notion of complacency is also the positive counterpart of trust. Trust is more often measured through a number of free response surveys that gauge it subjectively and are re-proposed to test participants across different scenarios/at different times [11]. Some collaborators called for the need to better assess trust in a nonsubjective frame, possibly accounting for number of re-engagements of ADS, gaps between ADS engagements, and time gaps to resume nondriving tasks. All three directions are currently under research (see, for instance, [42]), and collaborators called for unifying efforts from the research community on this front.
- Vigilance detection and countermeasures: Within the problem of assessing whether a driver is vigilant and responsive, the role that automation plays on mental workload needs to be considered [79]. Takeover tests only assess a somewhat heightened mental workload (a step change in mental workload and attention demand during the TOR), with drivers then assuming a more passive state before and after the transition in control authority. Incidentally, analyzing takeover performance is restricting the attention to what can be estimated to be less than 1% of the time the user spends in the vehicle, so that the before and after states need more careful attention. Furthermore, while the bulk of research has targeted the detection of low/high vigilance thresholds, few studies have assessed the role of countermeasures to re-engage the driver when low peaks of vigilance are detected. An interesting note in this regard comes from the trucking industry, where countermeasures in the form of incentive schemes are in place to retain attention levels of commercial drivers (e.g., a "five strikes" policy where a commercial driving license could be revoked for a period of time if inattentiveness and/or sudden hard braking events are detected too often). Driver-observation cameras are already widely deployed by commercial fleets.[3]

2. **HMI customization and mode-confusion avoidance**: While a host of HMI interfaces are currently being developed and tested, some collaborators pointed to the need to unify these studies into customizable and adjustable sets by each user. For example, a blind user may want to prioritize haptic feedback, while a deaf user may want to adjust specific coloring schemes in visual warnings. This is because with long-term exposure to new technologies, users may be receptive to mode-confusion (i.e., a decreased ability to correctly identify the system state and/or monitor its transitions). The possibility of high customization (perhaps guided by machine-learning applications that study the user performance over time) can help limit this possibility.
3. **Setting an aggregate measure for accuracy**: As pointed out in [14], time-related metrics for takeover performance alone do not necessarily reflect an effective control authority switch between the ADS and the human driver. For example, having a speedy response to the request to takeover by grabbing the steering wheel in 50% less time than other average participants does not imply that the driver is "safer" or "better" than the other test subjects. In particular, a driver could have a short time-to-input but then execute an imprecise takeover by drifting into the adjacent lane. This intuitive concept is reflected in the need to satisfy a combination of requirements R1 and R3. Aggregate measures for accuracy and effectiveness of the takeover need to be researched. A key obstacle has been that of ensuring that such metrics remain intuitive in their use, while maintaining operational bearing.
4. **Continue the work on new standards for AV signaling and communication:** This point shifts the discussion from within the vehicle to the outside world, and touches different strands of vehicle-to-everything (V2X) communication (including other vehicles, bystanders, and infrastructure). When conventional

[3] An example is Lytx Inc.'s recent agreements with Walmart Inc. and U.S. Xpress Enterprises. https://www.ttnews.com/articles/walmart-us-xpress-install-lytx-video-systems.

FIGURE 9. **Adapted version of NHTSA's estimated timeline for "five eras of safety."**

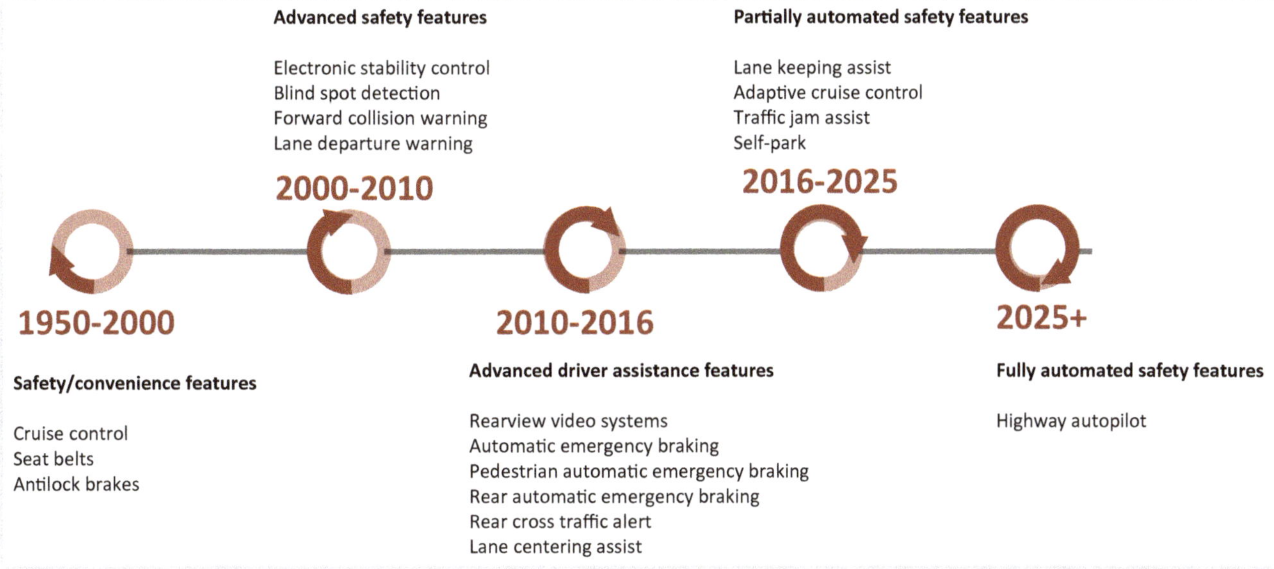

Adapted from NHTSA.

vehicles (and conventional drivers) are involved, an often implicit and tacit communication is established between drivers that helps us negotiate the right-of-way and share the road with other users [23]. Some authors refer to this behavior as an actual cooperation among drivers [38, 43] or coordination [57]. When at least one of those vehicles is replaced by an ADS-operated one, the balance in this implicit communication changes. For instance, when an L3 is considered, a driver might be present but not driving, giving a confused impression to other drivers if eye contact is not established, or possibly coming across as unresponsive (a typical trigger of frustration for other road users). It is then important to discuss whether there is a need (or a right, as discussed later) to notify other drivers that a vehicle is being operated by the ADS. Furthermore, should this need be established, standardization on how this communication would happen needs to be tackled. As a starting point, as suggested by [57] "an awareness and understanding of the effect and importance of movement gestures in day-to-day traffic interactions is needed for developers of autonomous vehicles to design forms of human-vehicle communication that are effective and scalable in multi-party interactions." From then, a list of important scenarios to tackle could be developed toward possible standardization efforts [72, 81]. SAE standard J2945 brought forward a list of possible requirements for messaging with "vulnerable road users" and can be considered a solid starting point to widen the range of users and scenarios considered.

Additional Topics

Progression to Full Autonomy

Table 1 and Figure 2 provided a depiction of what the AV market is shaping up to be. Collaborators discussed two approaches on the road to full autonomy: (i) a more gradual approach with further refinement of ADS technology that will progress through subsequent levels of automation, from semi-automated to fully automated, (ii) a "leap" to full autonomy that skips intermediate levels and passes from what we can term "advanced driver assistance" to full automation. Figure 9 provides an overview of the "five eras of safety" identified by National Highway Traffic Safety Administration (NHTSA),[4] which describes the "road to full automation" as a process "progressing through six levels of driver assistance."

While technological limitations shape the gradual development of highly automated systems, collaborators emphasized the importance of not forcing the six levels of automation onto a linear scale. In particular, the peculiarities of L3 in the vehicle interaction and need for collaboration with

[4] https://www.nhtsa.gov/technology-innovation/automated-vehicles-safety

the human driver point to what was referred to as a "step increase" from L2 and to the fact that L3 development poses a very different challenge than what L4 and L5 vehicles are tasked to become.

Concerns about L3 have focused on collaboration versus interaction for the monitoring function in L3, as explained next. While humans have a role within all level (more active, as drivers in L1 through L3, versus more passive for L4 and L5), L3 distinctively requires collaboration of the ADS and the human to accomplish the same task of environment monitoring. The vital monitoring role of the driver is no longer explicitly reflected within the SAE definition provided in Figure 1, although the notion of "fallback-ready user" indicates that the human operators must maintain active engagement. On one side, the user-ADS *interaction* framework for L4 (where we account for the transition from passenger to driver) presents a typical on/off performance of specific driving functions; on the other side, the *collaboration* at L3 must be concurrent to ensure the "readiness" for fallback. Ensuring that this collaboration can indeed happen is a challenge that some collaborators pointed to as "unsolvable," which is different in nature from those that other levels of automation are trying to overcome. The community needs to be wary of the inherent differences between the notions of semi-automation and full automation and to be careful in assessing the development across levels as progressive.

Between Morality and Social Value

Moral and ethical considerations related to fielding automated vehicles and to their interaction with humans from a social perspective were discussed with regard to the following unsettled domains:

- **Liability issues**: Collaborators of this report pointed to the difficulty in assigning responsibility for L3 operations, beyond the mere execution of the DDT. The interleaving in authority between the ADS and the human operator poses a challenge in assigning liability in the case of a crash or other incidents [13, 46]. This concern was also tied to the challenge of ensuring public acceptance of L3 systems and the need to have all users and stakeholders aware of liability demarcations (which are currently unset) between the vehicle operator and the manufacturer. Furthermore, liability considerations are not limited to the "user vs. manufacturer" debate within a single vehicle; in a scenario where a collision between an ADS-driven vehicle and a conventional vehicle ensue, the driver of the conventional vehicle at fault may appeal in its defense to the impossibility to adopt common driving norms and negotiation techniques with the "amorphous" automated vehicle. The opportunity regarding V2X communication with vulnerable road users discussed before touches on "inter-subjectivity" among different types of agents. The driver of the conventional vehicle may also appeal to the lack of proper policy or infrastructure in place to handle such situations, a "systemic effect" as mentioned next.
- **Shifting the focus beyond the driver - the importance of understanding systemic effects:** As the first fatal collisions ensue, larger systemic effects need to be considered in how they affect users. The work by [67] was the first to adopt a sociotechnical systems perspective to inform how federal and/or state policies and the regulatory climate (including standardization bodies) affect the on-road use of semi-automated system. The use of system theoretic process analysis (STPA) can help assessing how high-level stakeholders play a role within ensuring safe and efficient deployment of new technologies.
- **Public rights with regard to AVs**: The discussion on the need for standardized V2X signaling and communication hides within it a fundamental question about public rights - "Do people have a right to know that they are interacting with an AV?." Moreover, why and when should they know, and what implications does that knowledge carry? In a situation like that of drivers negotiating to occupy an intersection, communication serves a useful purpose, stemming from the consideration that for the next foreseeable future drivers will need to adapt to the coexistence of ADS-equipped and conventional vehicles. Similarly, a passerby who notices a car is stopped on the side of the road in an emergency could benefit from a signal that the vehicle does not need assistance, and that help is on the way. However, in nominal conditions of operation, signaling on the status of ADS-equipped vehicles could create a cognitive overload for drivers of conventional vehicles. Overload could, for instance, stem from the ignorance on the details of the specific ADS operations (e.g., "Is somebody in there?," "Am I on camera?," "Is it really driving itself?") and may not provide any safety benefit. Another side of the public rights discussion relates to requirements for public data reporting from the manufacturers on collisions, disengagements, and rides executed. Such is the case for the state of California, through the Department of Motor Vehicles testing program and the ride-hailing permit issued by the California Public Utilities Commission [16, 17, 18]. Different states across the U.S. are in the process of creating ad hoc legislation for automated vehicles [29, 52], but for now consensus has not been met on what public value such data offers, and on the rationale for this effort. Further discussion across the various stakeholders is needed to reach agreement.
- **Unintended uses**: Given the low maturity of high levels of automation, it is possible to foresee that humans will engage with this technology in some ways that we have not yet predicted. Naturalistic

studies beyond experimental settings and for extended time horizons need to look at how the technology engagement will play out in the long run. An example is the improper use of, for instance, an L2 vehicle where the user behaves as if it were L3, leading once more to liability issues and systemic effects on drivers' training and technology awareness.

Notes on Commercial Motor Vehicles

The unsettled topics and recommendations brought forward in this report equally apply to both personal transportation vehicles and commercial motor vehicles. Yet when similarities and differences between the two applications are considered, further unsettled topics can be identified as follows:

- **Refinement of ODD definition for L4 systems**: Within the trucking industry, specific applications for mining and construction work already deploy L4 systems. In these contexts, L4 motor vehicles mostly execute repetitive tasks on strictly defined paths. In other words, they operate within a very limited ODD with no plan to further augment such operational condition. This is a very different scenario than that of an L4 robo-taxi (e.g., Waymo's early rider program in Arizona) that currently operates within an ODD that gets further refined day by day to encompass more scenarios as they get validated. This difference might be indicative of a further refinement within what we consider an ODD and, in general, an L4 vehicle. This refinement, which could translate in novel terminology and/or honing of the levels of automation, would be tied to the scope of the ODD and the plans for its extension. At all effects, this would create a demarcation between those vehicles that are L4 due to technological limitations that prevent them from being L5, and those that are specifically targeting an L4 status with no plan to advance to L5.
- **Enhancement of levels of automation definition for "systems of vehicles"**: One of the most common applications of ADS for commercial motor vehicles is without doubt "platooning." Within the "master-slave" platooning context, we could face a situation in which the lead vehicle is being operated by a human driver, with a convoy of trailing ADS-operated followers with no drivers on board. From a system perspective, when we assess the entire convoy of vehicles it is not possible to assign an appropriate level of autonomy. This is because the lead may be considered L2 or L3, while the trailing vehicles could be considered L4. The same situation is encountered within "swarms" of automated vehicles, as in [31], that is, families of multiple automated vehicles of the same kind. The SAE J3016 levels of autonomy have an intrinsically single-system perspective and may need to be augmented by a multi-body system counterpart. While this may not seem a priority item, the SAE levels are permeating most of the AV literature, and it is important that the terminology and vocabulary set forth at the early stages be flexible to account for a perspective (i.e., that of multiple vehicles within a single system) that is likely to be often employed in the future. SAE might consider a separate standard or substandard that describes the different levels of automation that begin to appear in commercial vehicles.

Conclusions

Across the span of the SAE-defined Levels of Driving Automation, human drivers occupy a diverse range of responsibilities and authority on the vehicle movement and the monitoring of the outside environment. From both a technological and a regulatory perspective, there is a gap that divides lower levels of automation (L1 through L3) and higher levels of automation (L4 and L5). For those vehicles that require the cooperation between a human driver and the autonomous technology, it is important to ascertain the safety consequences of such a design choice and what the transition between automated driving and manual driving entails for the human driver, as well as for the surrounding traffic. In particular, the role of the user of an ADS-equipped vehicle has to be carefully defined before, during, and after the transition in control authority, and the problem of analyzing such transitions has to be more carefully tackled. Cross-disciplinary dialogue between the academic community, the manufacturers, and the regulatory agencies is needed to ensure that research efforts carry operational bearing, as well as standardization and certification are in place with clear guidelines for public users. This report overviewed a number of topics related to the qualification of the role of the human drivers within the vehicle as well as the quantification of their performance in taking over control and/or in remaining receptive to the status of the vehicle. A summary of the recommendations brought forward within the conversations with the collaborators is included on page 20.

SAE EDGE™ Research Reports

SAE EDGE™ Research Reports, like the present report on "Unsettled Issues Concerning Semi-Automated Vehicles: Safety and Human Interactions on the Road to Full Autonomy," are intended to push further out into still unsettled areas of technology of interest to the mobility industry. SAE launches these reports before attempting to form a joint working group, let alone a cooperative research program or standards committee.

SAE EDGE™ reports are intended to be quick, concise overviews of major unsettled areas where vital new technologies are emerging. By definition, an unsettled area is characterized more by confusion and controversy than established order. Early practitioners must confront an absence of agreement. Their challenge is often not to seize the high ground but to find common ground. These scouting reports from the frontiers of investigation are intended merely to begin the process of sorting through critical issues, contributing to a better understanding of key problems and providing helpful suggestions about possible next steps and avenues of investigation.

SAE EDGE™ Research Reports, therefore, are fundamentally distinct from the more formal working groups approach and far removed from the more mature CRP and standard's development process.

Next Steps for Semi-Automated Vehicles and the Role of Human Drivers

This publication should be considered only as a first step toward clarifying the issues around semi-automated vehicles and the role of human drivers. The intention behind this and other SAE EDGE™ Research Reports is to start a dialogue among interested parties on important industry-wide topics that require further attention. The expectation is that these explorations of unsettled areas of technology will lead to the formation of working groups and, ultimately, committees that can address and resolve the issues they raise, producing a framework for developing a common vocabulary of definitions, best practices, protocols, and standards needed to support continued progress toward safer and more innovative automotive products.

This SAE EDGE™ Research Report on semi-automated technology and its role toward full autonomy identifies the following key topics for further pursuit, both through continued informal discussions among industry practitioners and through more formal working groups:

- The need to further describe the role of human drivers across L3 and L4 vehicle, with particular attention to the transition from passenger to driver whenever a vehicle operates beyond the ODD
- The need to clarify the notion of "fallback-ready," including the associated metrics for quantifying takeover performance
- The needs for standardization of testing scenarios across multiple stakeholder including academic, regulatory entities, and manufacturers
- The need to investigate social acceptability and interactions among other road users
- The need to expand the current view of levels of autonomy when systems of vehicles are considered

Recommendations

The overall recommendations of this SAE EDGE™ report can be summarized as follows:

1. **Terminology**: The report detailed particular verbiage within the SAE J3016 standards that could benefit from further clarifications and possible revisions with regard to the following terms related to describing the role of a human driver within an ADS-equipped vehicle:
    - "Fallback-ready user" and its tie to the ODD beyond the targeted level of automation
    - "Evident failures" as pertaining to failures that may not trigger a notification from the ADS to the human user to regain control of the vehicle
    - "Not drivable" as pertaining to a vehicle where input HMIs are not present or no longer functioning

    Furthermore, the report brought forward the recommendation to look at the levels of automation from a system perspective, in the case in which more than one vehicle is being considered (e.g., platooning, or families of public transport AVs).
2. **Standardization**: The report detailed a series of applications within the research community that could benefit from more structured approaches to steer the efforts within a common direction. In particular, standardization should be discussed around the following themes:
    - Classification of metrics for the transition-to-manual problem, along with discussion on how to set measurement thresholds and how to establish the appropriate experimental apparatuses
    - Drafting of a common list of disengagement scenarios to tackle, with research effort providing a clear report (e.g., in the form of a code) of what scenarios are tackled from a list of pre-defined testing conditions
    - Drafting of a common list of independent variables to tackle, with research effort providing a clear report (e.g., in the form of a code) of what variable are tackled from a list of pre-defined ones (different HMI designs may be subsumed among those independent variables).
3. **New frontiers of research**: The report highlighted additional topics as "beyond unsettled" given the initial stage of research ongoing in those realms. The community is encouraged to tackle those in more depth, while regulatory and industry partners are needed to ensure the topics are accounted for from an operational context and given the appropriate amount of attention.
    - Defining aggregate measure for accuracy of the takeover performance, considering the low

operational bearing of most of the currently employed metrics.

- Expanding the time horizon of current studies, to account for behavioral adaptation, de-skilling, complacency buildup, and trust changes when long-term use of ADS-equipped vehicles is investigated. Longitudinal naturalistic studies can also help assessing possible misuse and unforeseen usage of those vehicles.
- Investigating systemic effects, as pertaining to the role that different stakeholders (including regulatory bodies at all levels) will have in shaping the safety culture that will surround ADS-equipped vehicles.
- HMI customization toward reducing the likelihood of mode confusion by different users.
- Public right and social acceptance, in relation to V2X communication between ADS-equipped vehicles and the surrounding environment.

Abbreviations/Definitions

ADAS - Advanced Driver-Assistance Systems

ADS - Automated Driving System

AV - Automated Vehicle

BAC - Blood Alcohol Concentration

DDT - Dynamic Driving Task

DUI - Driving Under the Influence

HMI - Human-Machine Interface

ODD - Operational Design Domain

OEDR - Object and Event Detection and Response

SJSU - San Jose State University

STPA - System Theoretic Process Analysis

TOR - Takeover Request

TTO - Time To Takeover

UAS - Unmanned Aircraft System

V2X - Vehicle-to-Everything

Acknowledgments

Recognition should go first to all the participants, many of who also provided feedback on the draft version of this publication. Without their input and initiative, this SAE EDGE™ Research Report would not have been possible.

Dr. Melissa Cefkin, Nissan Research Center
Dr. Alison Chaiken, Peloton Technology
Dr. Christian Janssen, Utrecht University
Angelos Lakrintis, Strategy Analytics
Dr. Mark Rosekind, Zoox; former NHTSA; former NTSB
Dr. Bernard C. Soriano, California Department of Motor Vehicles
Dr. Neville A. Stanton, University of Southampton
Dr. Ji Hyun Yang, Kookmin University

Additional thanks go to the following individuals who also had the opportunity to provide feedback on the draft version of this publication. Their contributions helped shape this publication.

Joseph Rodriguez
San Jose State University

The author of this document together with the SAE Team responsible for its creation joins in expressing our deepest appreciation to all the individuals mentioned above.

Francesca Favarò, PhD
San Jose State University

References

1. Alam, M.S., Ahmad, A., Khan, Z.A., Rafat, Y. et al., "A Bibliographical Review of Electrical Vehicles (xEVs) Standards," *SAE Int. J. Alt. Power.* 7(1):63-98, 2018, https://doi.org/10.4271/08-07-01-0005.
2. Anderson, J.M., Nidhi, K., Stanley, K.D., Sorensen, P. et al., *Autonomous Vehicle Technology: A Guide for Policymakers* (Santa Monica, CA: Rand Corporation, 2014).
3. Baek, S., Yun, H., and Yang, J.H., "How Do Humans Respond When Automated Request an Immediate Vehicle Control Take-over?," in *Proceedings of the 11th International Conference on Automotive User Interfaces and Interactive Vehicular Applications: Adjunct Proceedings*, Utrecht, The Netherlands, ACM, September 2019, 341-345.
4. Bahram, M., Aeberhard, M., and Wollherr, D., "Please take over! An Analysis and Strategy for a Driver Take over Request during Autonomous Driving," in *Intelligent Vehicles Symposium (IV)*, Seoul, South Korea, June 2015, IEEE, 913-919. https://doi.org/10.1109/IVS.2015.7225801
5. Banks, V.A. and Stanton, N.A., "Analysis of Driver Roles: Modelling the Changing Role of the Driver in Automated Driving Systems Using EAST," *Theoretical Issues in Ergonomics Science*, 20(3):284-300, 2017.
6. Banks, V.A., Eriksson, A., O'Donoghue, J., and Stanton, N.A., "Is Partially Automated Driving a Bad Idea? Observations from an On-Road Study," *Applied Ergonomics* 68:138-145, 2018.
7. Barber, P., *Applied Cognitive Psychology* (London: Routledge, 1988).

8. Bazilinskyy, P., and de Winter, J. (2015). Auditory Interfaces in Automated Driving: An International Survey. *PeerJ Computer Science*, 1, e13. https://doi.org/10.7717/peerj-cs.13

9. Bazilinskyy, P. and De Winter, J.C.F., "Analyzing Crowdsourced Ratings of Speech-Based Take-over Requests for Automated Driving," *Applied Ergonomics* 64:56-64, 2017, https://doi.org/10.1016/j.apergo.2017.05.001.

10. Beattie, D., Baillie, L., Halvey, M. and McCall, R., October 2014. What's around the Corner?: Enhancing Driver Awareness in Autonomous Vehicles via In-Vehicle Spatial Auditory Displays, "in *Proceedings of the 8th Nordic Conference on Human-Computer Interaction: Fun, Fast, Foundational*, Helsinki, Finland, ACM, 189-198.

11. Blanco, M., Atwood, J., Vasquez, H.M., Trimble, T.E., Fitchett, V.L., Radlbeck, J., Fitch, G.M., Russell, S.M., Green, C.A., Cullinane, B., and Morgan, J.F., Human Factors Evaluation of Level 2 and Level 3 Automated Driving Concepts (No. DOT HS 812 182), 2015.

12. Blommer, M., Curry, R., Swaminathan, R., Tijerina, L. et al., "Driver Brake vs. Steer Response to Sudden Forward Collision Scenario in Manual and Automated Driving Modes," *Transportation Research Part F: Traffic Psychology and Behaviour* 45:93-101, 2017.

13. Boeglin, J., "The Costs of Self-Driving Cars: Reconciling Freedom and Privacy with Tort Liability in Autonomous Vehicle Regulation," *Yale JL & Tech* 17:171, 2015.

14. Brumby, D.P., Davies, S.C., Janssen, C.P., and Grace, J.J., "Fast or Safe?: How Performance Objectives Determine Modality Output Choices While Interacting on the Move," in *Proceedings of the SIGCHI Conference on Human Factors in Computing Systems*, Vancouver, BC, Canada, ACM, May 2011, 473-482.

15. Caldwell, J.A., "Fatigue in Aviation," *Travel Medicine and Infectious Disease* 3(2):85-96, 2005.

16. California Department of Motor Vehicles, Article 3.7 - Autonomous Vehicles Testing. Title 13, Division 1, par. 227.02 (b), 2017a.

17. California Department of Motor Vehicles, Article 3.8 - Autonomous Vehicles Deployment. Title 13, Division 1, par. 228.06 (a), 2017b.

18. California Public Utilities Commission, Rulemaking 12-12-011, 2018, available at http://docs.cpuc.ca.gov/PublishedDocs/Published/G000/M215/K279/215279920.PDF.

19. Casner, S.M., Geven, R.W., Recker, M.P., and Schooler, J.W., "The Retention of Manual Flying Skills in the Automated Cockpit," *Human Factors* 56(8):1506-1516, 2014.

20. Clark, H. and Feng, J., "Age Differences in the Takeover of Vehicle Control and Engagement in Non-Driving-Related Activities in Simulated Driving with Conditional Automation," *Accident Analysis & Prevention* 106:468-479, 2017.

21. Clark, H., McLaughlin, A.C., Williams, B., and Feng, J., "Performance in Takeover and Characteristics of Non-Driving Related Tasks during Highly Automated Driving in Younger and Older Drivers," in *Proceedings of the Human Factors and Ergonomics Society Annual Meeting*, September 2017 (Vol. 61, No. 1, pp. 37-41). Sage CA: Los Angeles, CA: SAGE Publications.

22. Davtyan, K. and Favarò, F.M., "Effect of Phone Interface Modality on Drivers' Task Load Index in Conventional and Semi-Automated Vehicles," accepted for presentation at the in *3rd International Conference on Intelligent Human Systems Integration*, Modena, Italy, February 2020.

23. De Ceunynck, T., Polders, E., Daniels, S., Hermans, E. et al., "Road Safety Differences between Priority-Controlled Intersections and Right-Hand Priority Intersections: Behavioral Analysis of Vehicle-Vehicle Interactions," *Transportation Research Record* 2365(1):39-48, 2013.

24. Dogan, E., Rahal, M.C., Deborne, R., Delhomme, P. et al., "Transition of Control in a Partially Automated Vehicle: Effects of Anticipation and Non-Driving-Related Task Involvement," *Transportation Research Part F: Traffic Psychology and Behaviour* 46:205-215, 2017.

25. Eriksson, A., Banks, V.A., and Stanton, N.A., "Transition to Manual: Comparing Simulator with On-Road Control Transitions," *Accident Analysis & Prevention* 102:227-234, 2017.

26. Eriksson, A. and Stanton, N.A., "Driving Performance after Self-Regulated Control Transitions in Highly Automated Vehicles," *Human Factors* 59(8):1233-1248, 2017b.

27. Eriksson, A. and Stanton, N.A., "Takeover Time in Highly Automated Vehicles: Noncritical Transitions to and from Manual Control," *Human Factors* 59(4):689-705, 2017c.

28. Eurich, S., Favarò, F.M., and Agarwal, S., "Takeover Response Times Following Disengagements in Semi-Autonomous Vehicles," *Journal of Transportation Research*, 2020, under review.

29. Fagnant, D.J. and Kockelman, K., "Preparing a Nation for Autonomous Vehicles: Opportunities, Barriers and Policy Recommendations," *Transportation Research Part A: Policy and Practice* 77:167-181, 2015.

30. Favarò, F.M., Eurich, S., and Nader, N., "Autonomous Vehicles' Disengagements: Trends, Triggers, and Regulatory Limitations," *Accident Analysis & Prevention* 110:136-148, 2018a, https://doi.org/10.1016/j.aap.2017.11.001.

31. Favarò, F.M., Agarwal, S., Nader, N., and Mahmood, S., "Towards a Smart World: Hazard Levels for Monitoring of Autonomous Vehicles' Swarms," Mineta Transportation Institute, 2018, available at: https://transweb.sjsu.edu/research/Towards-Smart-World-Hazard-Levels-Monitoring-Autonomous-Vehicles%E2%80%99-Swarms.

32. Favarò, F.M., Seewald, P., Scholtes, M., and Eurich, S., "Quality of Control Takeover Following Disengagements in Semi-Automated Vehicles," *Transportation Research Part F: Traffic Psychology and Behaviour* 64:196-212, 2019a.

33. Favarò, F.M., Eurich, S., and Rizvi, S.S., ""Human" Problems in Semi-Autonomous Vehicles: Understanding Drivers' Reactions to Off-Nominal Scenarios," *International Journal of Human-Computer Interactions* 35(11):956-971, 2019b.

34. Favarò, F.M., Eurich, S., Agarwal, S., Rizvi, S., Nader, N., Mahmood, S., Akthar, S., Davtyan, K., and Derichs, M., "Impact of Smart Phones' Interaction Modality on Driving Performance for Conventional and Autonomous Vehicles," Mineta Transportation Institute Technical Report, editing stage, February 2020.

35. Forster, Y., Naujoks, F., Neukum, A., and Huestegge, L., "Driver Compliance to Takeover Requests with Different Auditory Outputs in Conditional Automation," *Accident Analysis & Prevention* 109:18-28, 2017.

36. Gold, C., Körber, M., Lechner, D., and Bengler, K., "Taking over Control from Highly Automated Vehicles in Complex Traffic Situations: The Role of Traffic Density," *Human Factors* 58(4):642-652, 2016.

37. Hayakawa, S., Yanagisawa, H., and Mita, S., "Ride Quality Evaluation for Autonomous Vehicles Using an Electrocardiogram," SAE Japan, Report 2011-08-0325, 2011.

38. Imbsweiler, J., Ruesch, M., Weinreuter, H., León, F.P., and Deml, B., "Cooperation Behaviour of Road Users in t-Intersections during Deadlock Situations," *Transportation Research Part F: Traffic Psychology and Behaviour* 58:665-677, 2018.

39. Izquierdo-Reyes, J., Ramirez-Mendoza, R.A., Bustamante-Bello, M.R., Pons-Rovira, J.L., and Gonzalez-Vargas, J.E., "Emotion Recognition for Semi-Autonomous Vehicles Framework," *International Journal on Interactive Design and Manufacturing (IJIDeM)* 12(4):1447-1454, 2018.

40. Janssen, C.P., Iqbal, S.T., Kun, A.L., and Donker, S.F., "Interrupted by my Car? Implications of Interruption and Interleaving Research for Automated Vehicles," *International Journal of Human-Computer Studies* 130:221-233, 2019.

41. Kim, H.J. and Yang, J.H., "Takeover Requests in Simulated Partially Autonomous Vehicles Considering Human Factors," *IEEE Transactions on Human-Machine Systems* 47(5):735-740, 2017.

42. Large, D.R., Burnett, G., Salanitri, D., Lawson, A., and Box, E., September 2019, "A Longitudinal Simulator Study to Explore Drivers' Behaviour in Level 3 Automated Vehicles," in *Proceedings of the 11th International Conference on Automotive User Interfaces and Interactive Vehicular Applications*, Utrecht, The Netherlands, ACM, 222-232.

43. Li, L. and Wang, F.Y., "Cooperative Driving at Blind Crossings Using Intervehicle Communication," *IEEE Transactions on Vehicular Technology* 55(6):1712-1724, 2006.

44. Liang, S.F., Lin, C.T., Wu, R.C., Chen, Y.C., Huang, T.Y., and Jung, T.P., "Monitoring Driver's Alertness Based on the Driving Performance Estimation and the EEG Power Spectrum Analysis," in *Proceedings of the IEEE Engineering in Medicine and Biology 27th Annual Conference*, Shanghai, China, September 1-4, 2005.

45. Louw, T. and Merat, N., "Are You in the Loop? Using Gaze Dispersion to Understand Driver Visual Attention during Vehicle Automation," *Transportation Research Part C: Emerging Technologies* 76:35-50, 2017.

46. Marchant, G.E. and Lindor, R.A., "The Coming Collision between Autonomous Vehicles and the Liability System," *Santa Clara Law Review* 52:1321, 2012.

47. Merat, N., Jamson, A.H., Lai, F.C., Daly, M., and Carsten, O.M., "Transition to Manual: Driver Behaviour when Resuming Control from a Highly Automated Vehicle," *Transportation Research Part F: Traffic Psychology and Behaviour* 27:274-282, 2014.

48. Michon, J.A., "A Critical View of Driver Behavior Models: What Do We Know, What Should We Do?" . In: Evans L., Schwing R.C., editors. *Human Behavior and Traffic Safety.* (New York, Plenum Press, 1985), 485-520.

49. Mok, B., Johns, M., Lee, K.J., Miller, D., Sirkin, D., Ive, P. and Ju, W., "Emergency, Automation Off: Unstructured Transition Timing for Distracted Drivers of Automated Vehicles," in *2015 IEEE 18th International Conference on Intelligent Transportation Systems*, Las Palmas, Spain, IEEE, September 2015, 2458-2464.

50. Murphey, Y., Kochhar, D.S., and Xie, Y., "Driver Workload in an Autonomous Vehicle," SAE Technical Paper 2019-01-0872, 2019, https://doi.org/10.4271/2019-01-0872.

51. Naujoks, F. and Neukum, A., "Timing of In-Vehicle Advisory Warnings Based on Cooperative Perception," in *Proceedings of the Human Factors and*

Ergonomics Society Europe Chapter Annual Meeting, Lisbon, Portugal, 2014, 1-13.

52. NCSL - National Conference of State Legislatures, 2019, "Database on Autonomous Vehicles Legislation," available from http://www.ncsl.org/research/transportation/autonomous-vehicles-legislative-database.aspx, query executed on 11/27/2019.
53. NHTSA - U.S. Department of Transportation, Federal Automated Vehicle Policy - Accelerating the Next Revolution in Roadway Safety, National Highway Traffic Safety Administration, September 2016.
54. NHTSA - U.S. Department of Transportation, Automated Driving Systems (ADS) 2.0 - A Vision for Safety, National Highway Traffic Safety Administration, 2017.
55. Mirae Asset Daewoo Co., Parts (Maintain) Industry Report, Mirae Asset Daewoo Co., Ltd., June 2017.
56. Rahman, H., Begum, S., and Ahmed, M.U., "Driver Monitoring in the Context of Autonomous Vehicle," in *SCAI*, Halmstad, Sweden, November 2015, 108-117.
57. Risto, M., Emmenegger, C., Vinkhuyzen, E., Cefkin, M., and Hollan, J., "Human-Vehicle Interfaces: The Power of Vehicle Movement Gestures in Human Road User Coordination," in *Proceedings of the Ninth International Driving Symposium on Human Factors in Driver Assessment, Training and Vehicle Design*, June 26-29, 2017, Iowa City, IA, Public Policy Center, University of Iowa, 2017, 186-192, https://doi.org/10.17077/drivingassessment.1633.
58. Rosekind, M.R., Boyd, J.N., Gregory, K.B., Glotzbach, S.F., and Blank, R.C., "Alertness management in 24/7 settings: lessons from aviation," *Occupational Medicine-Philadelphia* 17(2):247-260, 2002.
59. Rajendra, V. and Dehzangi, O., "Detection of Distraction under Naturalistic Driving Using Galvanic Skin Responses," in *2017 IEEE 14th International Conference on Wearable and Implantable Body Sensor Networks (BSN)*, Eindhoven, The Netherlands, May 2017, IEEE, 157-160.
60. Rizvi, S., Favarò, F.M., and Mahmood, S., "Participant Perception and HMI Preferences during Simulated AV Disengagements," in *International Conference on Applied Human Factors and Ergonomics*, Washington, DC, July 2019.
61. Rizvi, S., Favarò, F.M., and Nader, N., "Drivers' Situational Awareness and Impact of Phone Interface Modality in Conventional and Semi-Automated Vehicles," accepted for presentation at the in *3rd International Conference on Intelligent Human Systems Integration*, Modena, Italy, February 2020.
62. SAE International Surface Vehicle Recommended Practice, "Taxonomy and Definitions for Terms Related to Driving Automation Systems for On-Road Motor Vehicles," SAE Standard J3016, Rev. 2018.
63. Schipani, S.P., "An Evaluation of Operator Workload, during Partially-Autonomous Vehicle Operations," Army Research Lab, Aberdeen Proving Ground MD, available at: https://apps.dtic.mil/dtic/tr/fulltext/u2/a515361.pdf.
64. Sherdian, T.B., "Big Brother as Driver: New Demands and Problems for the Man at the Wheel," *Human Factors* 12:95-101, 1970.
65. Singh S, for NHTSA, "Critical Reasons for Crashes Investigated in the National Motor Vehicle Crash Causation Survey," DOT HS 812 115, US Department of Transportation, 2015, available from https://crashstats.nhtsa.dot.gov/Api/Public/ViewPublication/812115
66. Stanton, N.A. and Marsden, P., "From Fly-by-Wire to Drive-by-Wire: Safety Implications of Automation in Vehicles," *Safety Science* 24(1):35-49, 1996.
67. Stanton, N.A., Salmon, P.M., Walker, G.H., and Stanton, M., "Models and Methods for Collision Analysis: A Comparison Study Based on the Uber Collision with a Pedestrian," *Safety Science* 120:117-128, 2019.
68. Statista - US ADAS unit production volume 2014 to 2021 & US passenger car production 2017, accessed on 10/2/2018, 2018, available at https://www.statista.com/statistics/442609/production-volume-of-advanced-driver-assistance-systems-in-the-us/
69. Tunnell, J., Asher, Z.D., Pasricha, S., and Bradley, T.H., "Toward Improving Vehicle Fuel Economy with ADAS," *SAE Int. J. CAV* 1(2):81-92, 2018, https://doi.org/10.4271/12-01-02-0005.
70. van der Heiden, R., Iqbal, S.T., and Janssen, C.P., "Priming Drivers before Handover in Semi-Autonomous Cars," in *Proceedings of the 2017 CHI Conference on Human Factors in Computing Systems*, ACM, Denver, CO, May 2017, 392-404.
71. van der Heiden, R.M., Janssen, C.P., Donker, S.F., Hardeman, L.E., Mans, K., and Kenemans, J.L. (2018). Susceptibility to Audio Signals during Autonomous Driving. *PloS One*, 13(8), e0201963.
72. Vinkhuyzen, E. and Cefkin, M., "Developing Socially Acceptable Autonomous Vehicles," in *Ethnographic Praxis in Industry Conference Proceedings*, Minneapolis, MN, 2016 (Vol. 2016, No. 1, pp. 522-534).
73. Walker, G.H., Stanton, N.A., and Salmon, P.M., *Human Factors in Automotive Engineering and Technology* (Aldershot, UK: Ashgate, 2015).
74. Waterhouse, J., Edwards, B., Atkinson, G., Reilly, T. et al., "Occupational Factors in Pilot Mental Health: Sleep Loss, Jet Lag, and Shift Work," In: Hubbard T., Bor R., editors. *Aviation Mental Health: Psychological Implications for Air Transportation.* (London, Taylor and Francis Group, 2016), 255-255.

75. Wester, A.E., Böcker, K.B.E., Volkerts, E.R., Verster, J.C., and Kenemans, J.L., "Event-Related Potentials and Secondary Task Performance during Simulated Driving," *Accident Analysis & Prevention* 40(1):1-7, 2008.

76. Wickens, C.D., Gordon, S.E., Liu, Y., and Lee, J., *An Introduction to Human Factors Engineering* (Charleston, SC, 1998).

77. Yang, J.H., Mao, Z.H., Tijerina, L., Pilutti, T. et al., "Detection of Driver Fatigue Caused by Sleep Deprivation," *IEEE Transactions on Systems, Man, and Cybernetics-Part A: Systems and Humans* 39(4):694-705, 2009.

78. Yang, J.H., Han, J., and Park, J., "Toward Defining Driving Automation from a Human-Centered Perspective," in *Proceedings of the 2017 CHI Conference Extended Abstracts on Human Factors in Computing Systems*, Denver, CO, May 2017, ACM, 2248-2254.

79. Young, M.S. and Stanton, N.A., "Malleable Attentional Resources Theory: A New Explanation for the Effects of Mental Underload on Performance," *Human Factors* 44(3):365-375, 2002.

80. Zeeb, K., Buchner, A., and Schrauf, M., "Is Takeover Time All That Matters? The Impact of Visual-Cognitive Load on Driver Takeover Quality after Conditionally Automated Driving," *Accident Analysis & Prevention* 92:230-239, 2016.

81. Zhang, J., Vinkhuyzen, E., and Cefkin, M., "Evaluation of an Autonomous Vehicle External Communication System Concept: A Survey Study," in *International Conference on Applied Human Factors and Ergonomics*, Springer, Cham, July 2017, 650-661.

Contact Information

EDGEresearch@sae.org.

The authors of this document together with the SAE Team responsible for its creation join in expressing our deepest appreciation to all of the individuals who contributed to its development.